María Del Rosario Vázquez Pérez

The cuisenaire strips

María Del Rosario Vázquez Pérez

The cuisenaire strips

A strategy for teaching mathematics in preschool education

ScienciaScripts

Imprint

Cover image: www.ingimage.com

This book is a translation from the original published under ISBN 978-613-9-43488-6.

Publisher:
Sciencia Scripts
is a trademark of
Dodo Books Indian Ocean Ltd. and OmniScriptum S.R.L publishing group

120 High Road, East Finchley, London, N2 9ED, United Kingdom
Str. Armeneasca 28/1, office 1, Chisinau MD-2012, Republic of Moldova, Europe
Managing Directors: Ieva Konstantinova, Victoria Ursu
info@omniscriptum.com

Printed at: see last page
ISBN: 978-620-8-40900-5

CUISENAIRE'S RULLETTES; A STRATEGY FOR TEACHING MATHEMATICS IN PRE-SCHOOL EDUCATION

MARIA DEL ROSARIO VAZQUEZ PEREZ

INTRODUCTION

As a result of the developmental processes and experiences that children have when interacting with their environment, they develop numerical, spatial and temporal notions that allow them to advance in the construction of more complex mathematical notions. from a very early age and according to the stimuli they receive, they can establish relations of equivalence, equality and inequality (more, less or equal quantity); they realise that "adding makes more" and "taking away makes less", and distinguish between large and small objects. their judgements appear to be genuinely quantitative and they express them in various ways in situations of their everyday life.

The natural, social and cultural environment in which young children develop provides them with experiences that spontaneously lead them to carry out counting activities, which are a basic tool for mathematical thinking. In their games or in other activities, they separate objects, give out sweets or toys; when they carry out these actions, and although they are not aware of it, they begin to implicitly and incipiently put into practice the principles of counting: one-to-one correspondence, stable order, cardinality, irrelevance of order and abstraction.

During preschool education, activities through play and problem solving contribute to the use of counting principles (numerical abstraction) and counting skills (beginning numerical reasoning) so that children gradually build the concept and meaning of number (pep 2011).

The diversity of situations proposed to the pupils makes them increasingly able, for example, to count the elements in an array or collection, and to represent in some way the number of elements (numerical abstraction); they will be able to infer that the numerical value of a set of objects does not change just by dispersing the objects, but changes - increases or decreases in value - when one or more elements are added to or removed from the set or collection. Thus, abstraction skills help them to establish values and numerical reasoning allows them to make inferences about and operate with established numerical values.

At pre-school level we have the obligation as teachers to look for and create different strategies to get children to develop logical mathematical thinking; as an educator I have discovered in the use of Cuisenaire Rulers a method that allows us to establish numerical relationships in a geometric way, so these relationships will be much more visual and manipulative for children.

WHY USE NUMBER STRIPS IN PRE-SCHOOL MATHEMATICS LEARNING?

With the number strips, mathematics reaches an extraordinary sensory level. Experiences should be provided in a way that activates as many senses as possible, and this is something that is achieved with this material, turning a very abstract area such as mathematics into something concrete that the child can manipulate and visualise in a clear way, respecting and facilitating their process of abstraction.

It is important to set up challenging activities where children will make their own discoveries and generate their own hypotheses and answers based on their manipulation and research. we should not make the mistake of starting at the end and presenting the mathematical concepts directly with the rulers, but let the children themselves come to these conclusions through their own work and encouraged by some key questions that we can ask them to guide them in their discoveries.

Within the teaching practice and the approach of challenging activities, this question arises: what importance do we teachers give to the use of cuisenaire rulers as an effective didactic material in the teaching of mathematics in pre-school education?

At pre-school level, the aim is to contribute to the transformation of educational practices in the classroom, so that children have interesting and challenging learning opportunities at all times that promote the achievement of fundamental learning, always based on the knowledge and skills they already possess.

As an educator, moving towards this goal has been a learning process that involves trying innovative ways of working with students, making mistakes, reflecting, evaluating, trying again and discovering in these attempts at change that children have multiple capacities and that it is possible to propose activities that bring them to the surface and that demand a cognitive challenge from them.

Promoting the achievement of knowledge in diverse situations and contexts has to do with the learning processes that I make possible as an educator with the activities that I propose and through the teaching intervention.

This article aims to recover evidence and express my experience in educational practices, which allow me to make a judgement on the usefulness and impact of Cuisenaire's rulers as tools that strengthen the processes of teaching mathematics in preschool students, and also to reiterate that the use of rulers as didactic tools in the teaching of mathematics is a favourable way to develop mathematical logic skills in preschool children.

THEORETICAL BACKGROUND

A question suggested by Irma Fuenlabrada that can guide the discussion is: are children in their transit through pre-school education being given the possibility to develop learning correlated with knowledge of number?

One way to find out is whether, when faced with different situations or problems, children have opportunities to perform the following actions linked to reasoning:

a) . To look for a solution to the situation; that is, if they show an attitude of security and certainty as the thinking subjects that they are.

b) . To understand the meaning of numerical data in the context of the problem, i.e. to show their mathematical thinking.

c) . Choose, from the knowledge they have learnt (numbers, their representation, counting, additive relationships, etc.), the one they can use to solve the situation.

d) . Use this knowledge fluently to solve (skills and abilities) the given situation.

irma fuenlabrada suggests that we do a small exploration in the group and what if it turns out that when we give the pupils a problem that involves adding, gathering, taking away, matching, comparing and distributing objects, the children wait for the indications to proceed, we should make the following assessment:

- If the children are in the first grade of pre-school, there is no problem, we still have two more years to develop learning, not only about numerical knowledge but also about how to act in the face of what they do not know. but without losing sight of the fact that to achieve this it is essential to systematically allow the children to use their own resources to find ways of solving the various mathematical situations that are proposed to them.

The aim of the pre-school level is for teachers to foster the development of learning in their pupils, some of the aims of the pre-school level are:

1. Use mathematical reasoning in a variety of situations that require the use of counting and first numbers.

2. Understand the relationships between data in a problem and use one's own procedures to solve them.

Working with ruler strips helps us to develop the use of mathematical reasoning, generates mental agility, contributes to the formulation and resolution of problems that involve adding, gathering, taking away, equalling, comparing and distributing objects, and provides an understanding of the four basic operations (addition, subtraction, multiplication and division at primary level).

The development of reasoning skills in pre-school students is fostered when they perform actions that allow them to understand a problem, reflect on what is being sought, estimate possible results, look for different ways to solve it, compare results, express ideas and explanations and confront them with their peers. this does not mean rushing the formal learning of mathematics, but rather fostering the forms of mathematical thinking that children possess towards the achievement of the competencies that are the foundation of more advanced knowledge that they will build throughout their schooling. (pep 2011).

At the pre-school level, several phases are worked on in the children's action with the cuisenaire rulers, namely:

1. Spontaneous games, where pupils freely manipulate the material, without adult intervention;

2. Empirical search: in this phase children develop their activity with a certain intention such as building trains, decomposition of strings, pairs (size comparison),

3. Systematisation and mastery of the structures: through experience, the child becomes familiar with the possibilities of the strips and knows their structural aspects, and gradually frees himself from the material and is able to create mathematical facts by himself: adding, gathering, removing, equalling, comparing and distributing objects. at this point he is invited to describe numerical relations, he does so without difficulty and feels a real pleasure in doing so, seeking new possibilities of mathematical expression that enrich his knowledge.

The idea of starting with free play is that they begin to recognise the rulers through manipulation and exploration, considering that it is through action that the child constructs mathematics; starting from the pupil's own knowledge.

Particularly in pre-school it is important for the child to visualise, play with the strips and use his or her imagination to build with the strips.

In pre-school, children can work on operations such as addition, subtraction, multiplication and division, but empirically (which in pre-school language means adding, subtracting, equalling, joining, comparing and dividing), to the extent that the teacher guides and accompanies the work with rulers without naming each operation, ("strengthening numerical thinking through cuisenaire's rulers").

It is important that the activities are presented to the child in such a way that he/she arrives at the concepts through discovery and that we are not the ones who give him/her the solution before starting the challenge.

I consider it valid to introduce students to the material when they do not know it, let them explore it, let them identify shape, size, colour and discover freely how to use it, so that, in accordance with the activities with educational intent, they can handle it and develop mathematical learning.

ACTIVITIES WITH THE CUISENAIRE RULLES AT PRE-SCHOOL LEVEL

1. Free play.

With time and constant use of the number strips, the child will internalise the numerical value of each one of them and will establish relationships between the different pieces and although he/she does not know that he/she is carrying out mathematical operations, he/she will be applying relations of equivalence and quantity and putting logical reasoning into practice to solve situations that require him/her to add, remove or equal quantities.

It is recommended, as with all materials used for the first time, that we allow the children to play freely with the rulers for several sessions, so that they become familiar with the material.

2. Sorting strips

Place the strips on the table and ask the pupils to put "together those that go together" without giving them the answer of doing it by colour and size, leaving it up to them to come up with this reasoning and the concept of CLASSIFICATION.

3. Discover the power strip

The pupils take three strips, for example, red, green and yellow, hide them in their hands and put them behind their backs, the teacher asks them to show her one strip naming it by colour, for example: "show me the red strip", then put it back together with the other three, hide them and ask her for another colour, just by touch they have to identify the strip she is asking them for, thus associating a length with each colour.

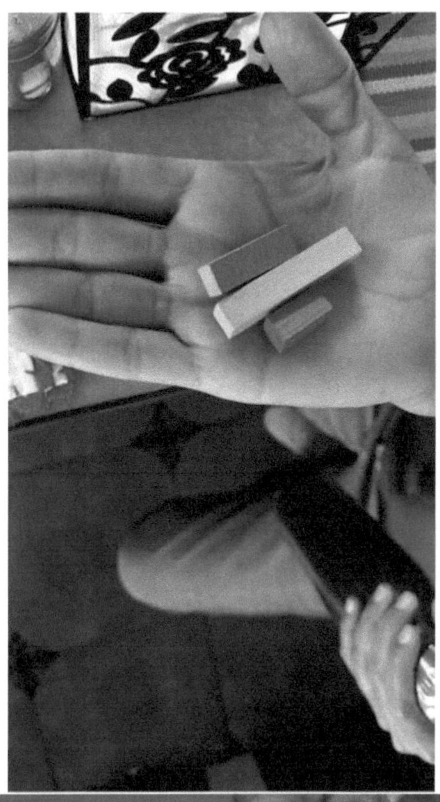

4. Ladders from smallest to largest (increasing order) and from largest to smallest (decreasing order)

Each child is given 1 strip of each colour and is instructed to form them from the white strip to the orange strip in a free way, with the intention that they identify the logical sequence of size and if there is any error they reason and manage to solve it. Once they have solved the ladder in ascending order, they are instructed to do it in descending order.

5. Pyramid

The children are given 2 strips of each colour per child, and are instructed to first form a ladder, starting from the white strip to the orange strip, in ascending order, so that once they have finished, they can make a ladder starting from the orange strip and ending at the white strip, in descending order. Mention these concepts so that the children can make them their own.

6. Find out which one is missing

Due to the complexity of the activity, it will be carried out individually so that the group is the participant's spectator.

Once finished, ask the child to close his eyes and remove one of the strips, put all the strips together so that it is not obvious where we have taken it from and ask him to open his eyes and discover which strip is missing, give him time to observe and discover it on his own, if necessary, the support of a classmate can be requested.

Variation: For 4 year olds, we can show them the power strip that we have removed so that they only discover the place it occupies on the stairs, as they become familiar with the proposal, without them knowing which power strip has been removed.

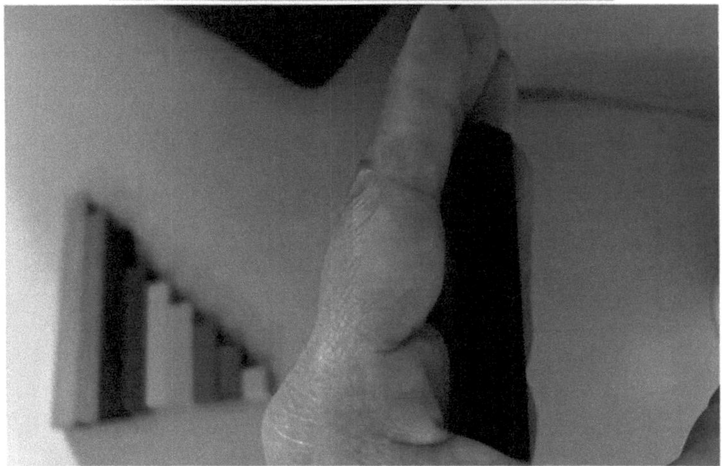

7. Establish numerical relationships between them

Make two vertical ladders, one increasing and one decreasing, facing each other. Ask them to join them together to form a square with the two of them, without any extra pieces or gaps.

Once the task has been completed, reflect on what has been observed:

*The geometric pattern is repeated in reverse.

*The repeating strip (the 5) is the one that marks this change in the pattern.

*All pairs of strips have a value of 10

8. Game of cinquillo

In this game the children will work with the number series from 1 to 10 in both ascending and descending directions.

We will play as a group with a team of 4 players and 40 strips are needed (four of each colour), all the strips are distributed arbitrarily among the four children, the first player places a yellow strip in the centre of the table, if there are no yellow strips it is the turn of the next player.

once the yellow strip has been placed, the next player must place a pink strip or a light green strip to build a staircase from the yellow strip. if he does not have one, he can place another yellow strip in another area of the table to build another staircase. if he does not have a yellow strip either, he passes the turn without placing any strip.

The next player has to either place a rail immediately above or below the ones at the ends of the train or start a new ladder with the yellow rail, the first player to run out of rails wins.

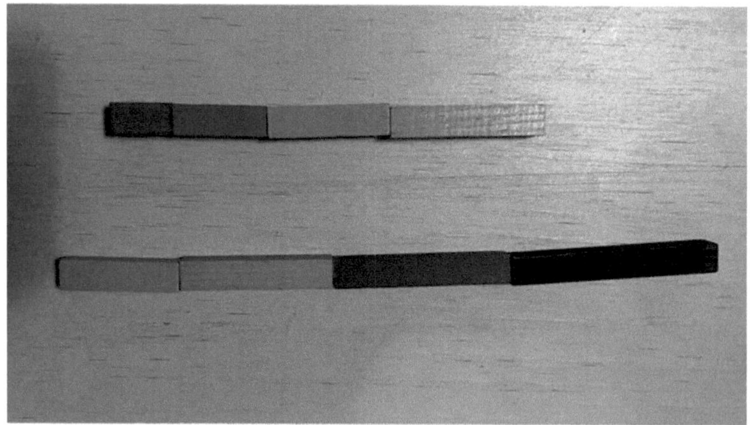

9. Establish equivalences based on the unit

We take it in turns to guess how many squares of the unit each strip corresponds to.

We can use the following question for the child to solve, How many white strips fit in the strip: orange, blue, green or pink, with the intention of arriving at the reasoning of what equivalence means, which is nothing more than equality in the value in this case of the strips.

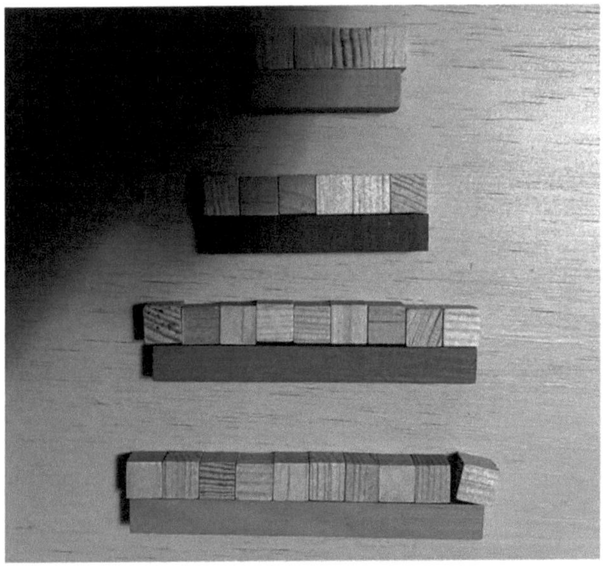

*After working with the white strips, find the equivalence with the other strips, check that each strip is one more than the following strip: red is white plus one, green is red plus one, yellow is pink plus one ....

*This activity can be done with any other number, not just 10: how many strings are equivalent to 5, 6, ..., but given that our number system is decimal, it is convenient to work a lot on changes in 10.

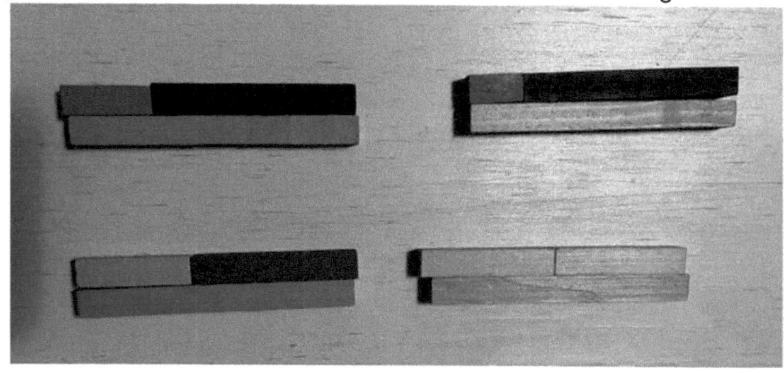

*Make equivalences using 3 or more power strips, e.g. which ones form the yellow, black, green power strip?

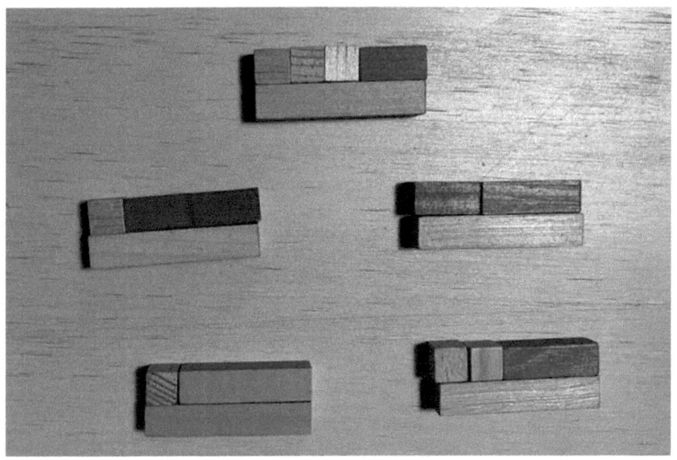

*Make cubes: with strips of the same colour, or mixing them, give each team a handful of strips from yellow to white and ask them to make a cube using all the strips they have.

*play to change strips between two players, give a handful of strips to each player and let them change them between them, it is important that the changes are always equivalent, for example, change one of 7 for one of 2 and another of 5.

Play the coloured snake: after placing a handful of strips in a row, making little paths, snakes, figures, we are going to calculate how many there are, to do this they are all placed in a row, 1 orange strip (10) is placed in parallel, which will be the basis for forming other snakes, but we can start from another of a lower value and increase until we reach 20 for example, using 2 orange strips as a base.

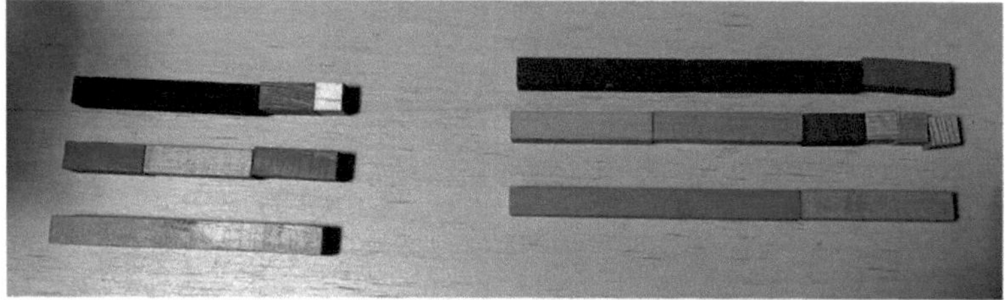

10.Capture strips

The game is played between two children or two teams seated opposite each other, each team has in front of them, placed in a row, a strip of each colour in order from highest to lowest, the dice is rolled in turn and according to the numbering the strip or strips are taken from the opposing team, the winner is the one who captures all the strips of the opposing team.

The pupils can play with the first six strips and with only one die, and the number can be increased until two dice are used.

11. Bingo with strips

Each child is given a board containing the colours of some strips and is given tokens so that as the numbers from 1 to 10 are mentioned, the children associate the colour with the corresponding strip. For example: five, and whoever has a yellow colour on their board gets a token.

Whoever completes their board first wins.

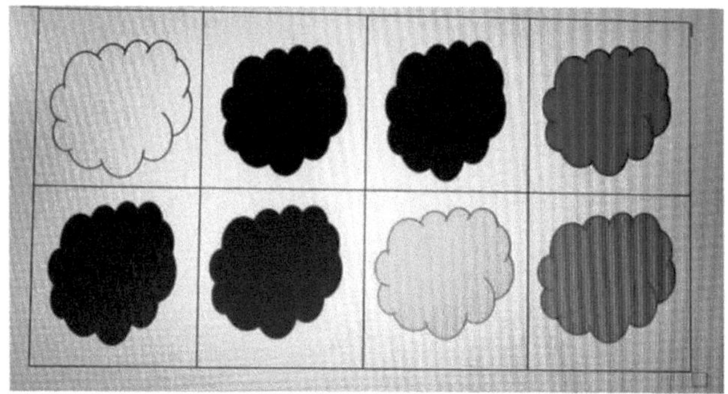

12. Balancing of strips

The group can be divided into two teams, a box of strips is placed in the centre and in this game a numerical die is used, the first player throws the die and according to the number that falls the team places the corresponding strip in front of it, it is the turn of the opposing team to do the same action, and so each member continues to participate until a tower is formed. The winner is the team that manages to make its tower higher or keep it in balance.

13. Comparison of numbers: greater than / less than, equal to

The comparison of numbers is extremely visual with the strips, just zoom in on each other and compare their sizes.

In this activity we will use the greater than, less than and equal signs, so that the children become familiar with them and acquire the reasoning of inequality and equality.

14. Solve problems involving adding (addition)

To solve problems involving addition, we place the strips next to each other and underneath we put the result.

Note: The image with the + and - signs are for reference only, as they are not used in pre-school.

We will also be able to solve problems with more than two digits and with the corresponding strings to represent the result.

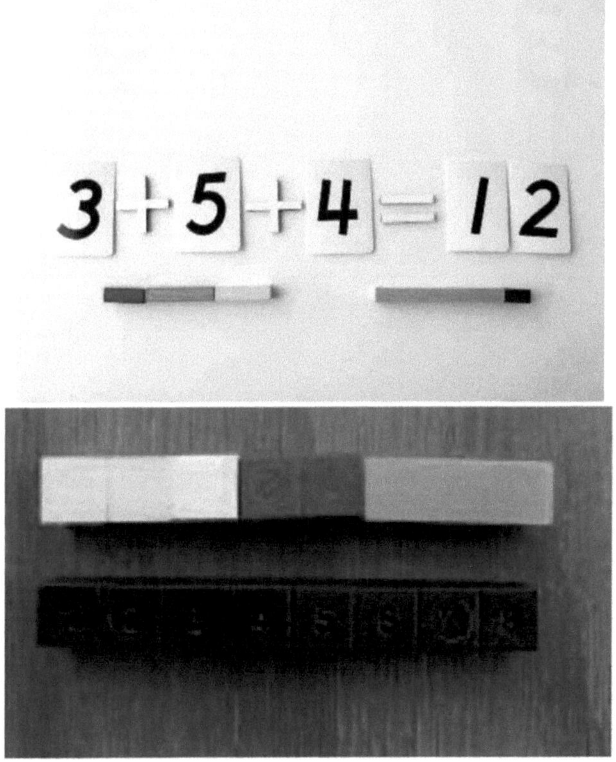

15. The hidden power strip

We can play to find out which is the ruler that corresponds to the empty space and that, when joined to the ruler that does appear, gives the result that is obtained at the end.

16. Composition and decomposition of sets

This game consists of solving problems that involve adding, we can play to see how many sets of strips we can build that give us the same result (equivalence), in this way we will also be working on the composition of numbers and complementary numbers.

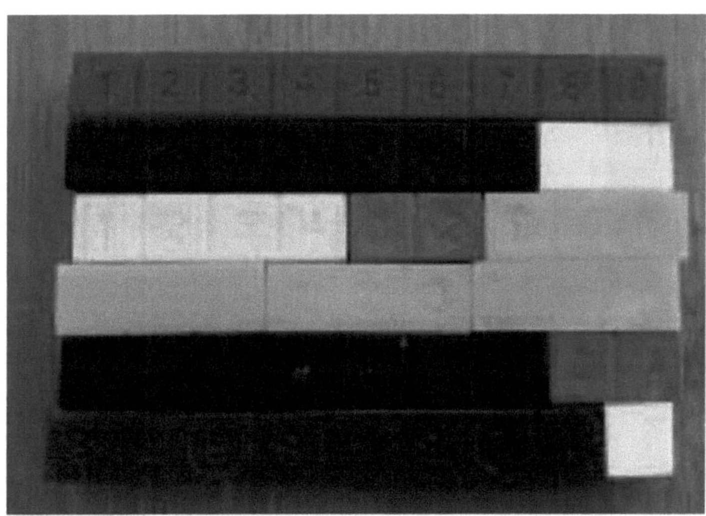

17. Commutative property of sets (changing the order)

With the strips we can see very visually that no matter in which order the strips are placed, the result will always be the same.

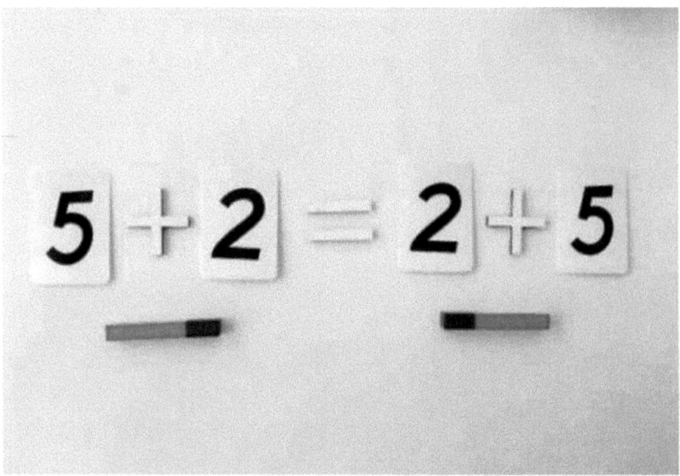

5 + 2 = 2 + 5

18. Associative property of sets

When we join more than two strips we can associate them in any way we want, the result will not change.

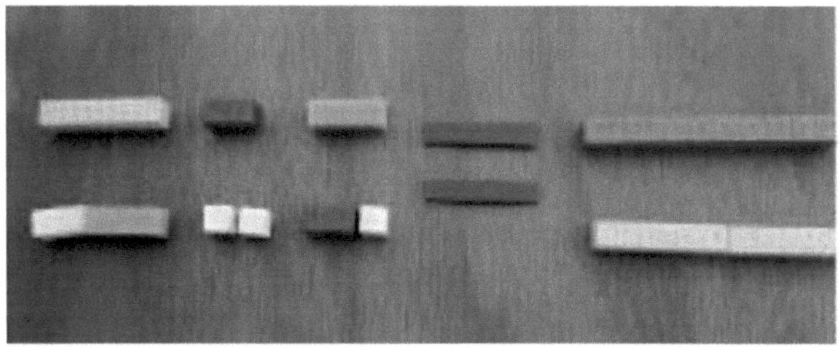

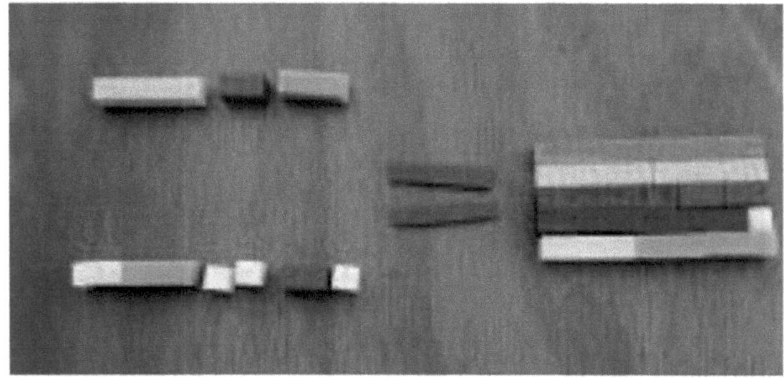

19. Actions involving the removal of

To solve problems that involve removing two numbers represented by strips, we will put the larger strip on top and the one we have to remove below it, we will approach it in the following way: if we have this strip and we remove what this smaller one occupies, how much would we have left? then we look for a strip that occupies the same as the space we have left and that would be the result of the problem.

6
2
6
2

20. Discover the hidden power strip

We can play to find out which is the ruler that corresponds to the empty space and that, when joined to the ruler that does appear, gives the result that is obtained at the end.

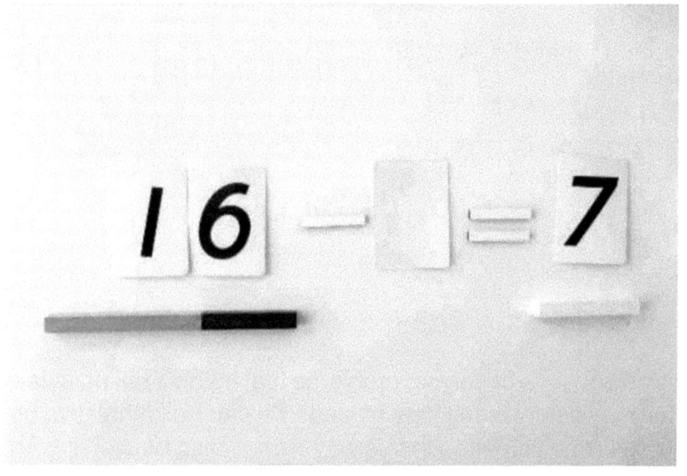

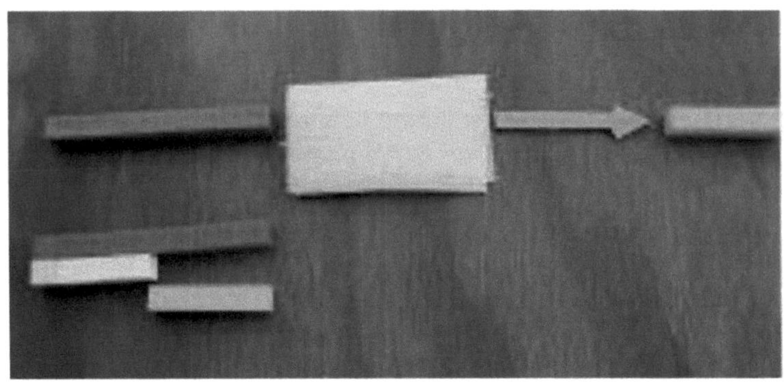

21. Search for sets with the same result

This game consists of solving problems that involve taking away, we can play to see how many sets of strips we can build that give us the same result (equivalence), in this way we will also be working on the decomposition of numbers and complementary numbers.

22. The numbers 11 to 20

With the rullettes, equivalences greater than 10 will be made and that maximum reach the number 20, as it is very visual that 11 is 10 and 1, 12 is 10 and 2, 13 is 10 and 3... but once the child has acquired this notion, these quantities can also be composed with different rullettes, 11 is 8 and 3, 12 is 6 and 6, 13 is 8 and 5.

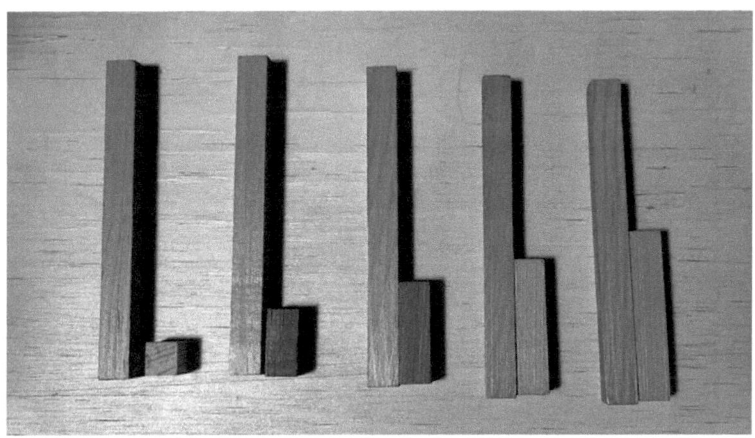

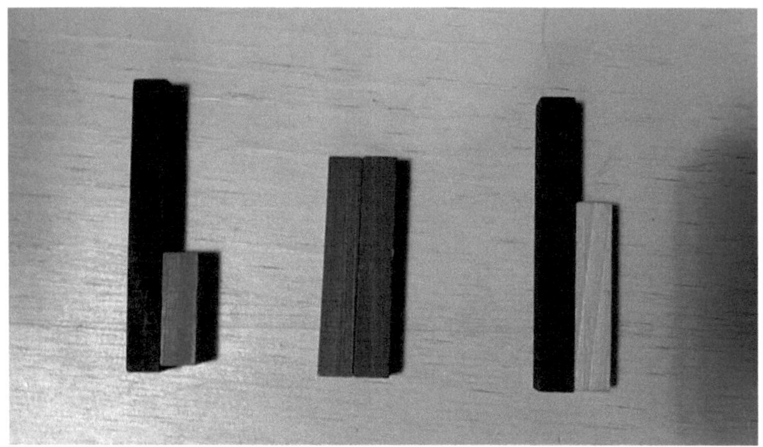

23. Solving mathematical problems involving adding, subtracting and equaling.

* I have 3 popsicles and my mum brought me 4 from the shop, how many popsicles do I have in total?

* I have 6 pallets, if I give 3 as a gift, how many do I get?

*My mother gave me two fish tanks, in one I have 2 fish and in the other 5, how many fish do I need so that the two tanks have the same amount of fish?

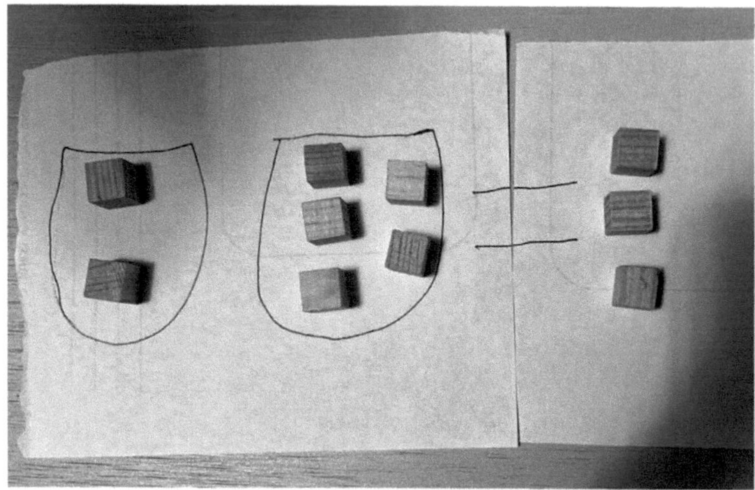

Note: In the images you can see symbols such as +, -, <, > and =, this does not mean that we should or can apply them in pre-school, it is only a reference of the operation we are carrying out, such as adding, taking away, equalling, making differences between greater and lesser quantity, etc.

CONCLUSIONS

The use of Cuisenaire's Rulers has a wide variety of applications as it allows pre-school students to develop their abstract thinking (the ability to change, at will, from one situation to another, to break down the whole into parts and to analyse simultaneously different aspects of the same reality) and to construct their own learning, there are different levels of development of mathematical logical thinking in children, For this reason, it is necessary to make use of these materials to support students in reaching the "zone of proximal development", and I believe that if these types of activities are unknown or have not been carried out in some schools, it is because they do not have these types of resources or, even if they do, they are not given the proper use and the educational intention that they deserve.

Vygotsky defined the zones of proximal development as the distance from "the level of the child's actual development as it can be determined from independent problem solving and the higher level of potential development as it is determined by problem solving under adult guidance or in collaboration with more capable peers". (Wertsch, james v.: Vygotsky y la formación social de la mente. editorial Paidós).

So it is not impossible to work with rulers at pre-school level starting from second grade, as we know beforehand that the activities are graduated according to the age, characteristics and needs of the group, the mental barriers may have the teacher and it depends on him whether the children have them or eliminate them.

Working with the Cuisenaire Rulers also requires teachers to be permanently prepared and to assume the role of learning coordinator for their students, as it generates an interesting culture where students build their own concepts and share them with their peers.

From my point of view, working with the rulers makes working with mathematics much easier and more fun, as the pupils interact with the material and at the same time build new knowledge.

The implementation of this material is very effective for any type of mathematical operation that involves deep reasoning to solve situations, but above all where the student can manipulate to have a more meaningful learning, they are a material, which provides the opportunity to develop mathematical skills from the game, manipulation and experimentation.

BIBLIOGRAPHICAL REFERENCES

Secretaría de educación pública, (2017), Aprendizajes clave para la educación integral, plan y programas de estudio, orientaciones didácticas y sugerencias de evaluación.

Secretaria de educación pública, (2004) Programa de educación preescolar.

Secretaria de educación pública, (2011) Programas de estudio 2011. guía para la educadora.

Fuenlabrada irma, up to 100?...no! and the accounts?...not then either.... what? sep. 2009

María Fanny Nava , Luz Marina Rodríguez , Magda Patricia Romero , María Elvira Vargas. Article: "Strengthening numerical thinking through Cuisenaire's Rulers".

Compilation of some images from the internet.

Table of Contents

Printed by Books on Demand GmbH, Norderstedt / Germany